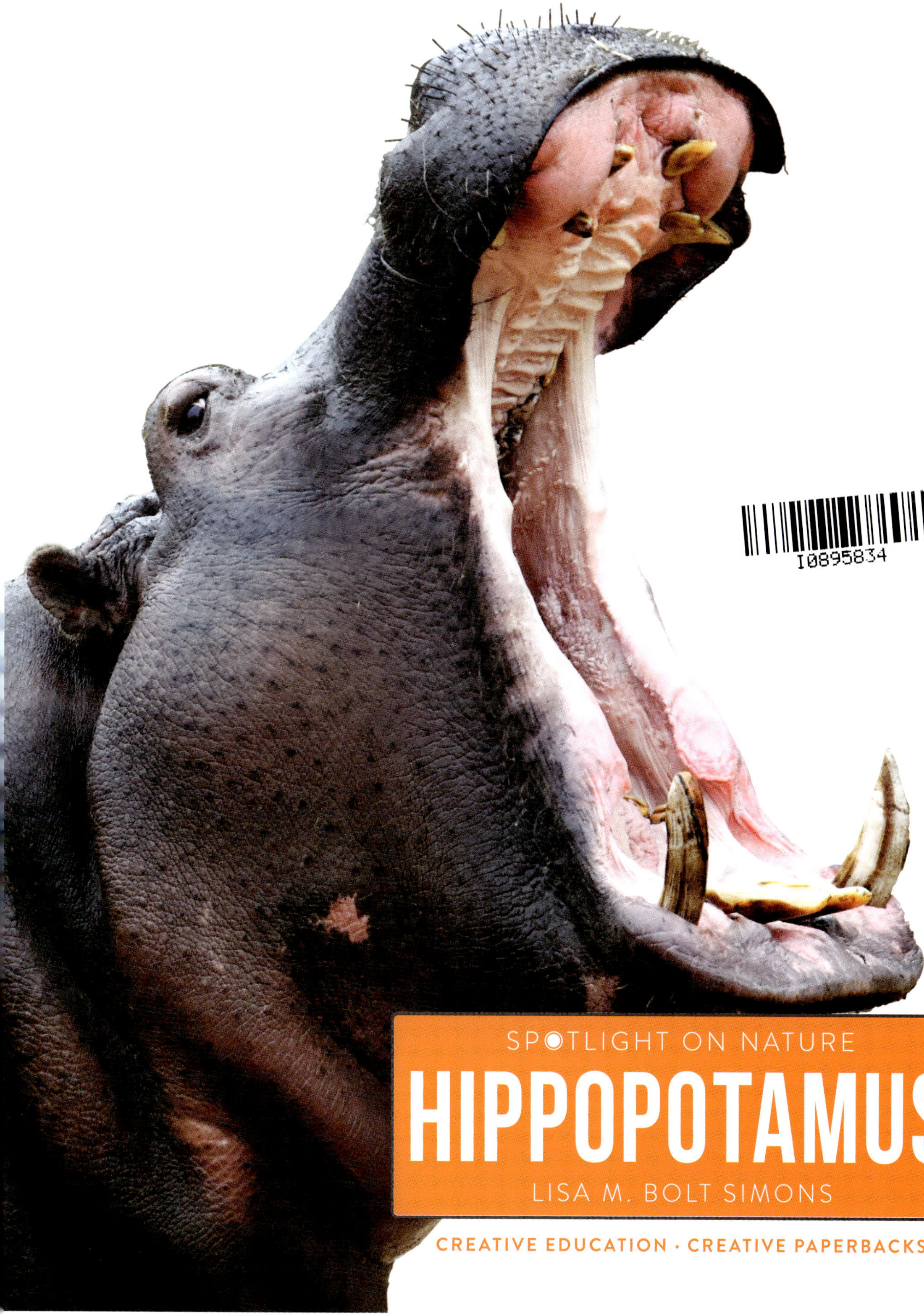

SPOTLIGHT ON NATURE

HIPPOPOTAMUS

LISA M. BOLT SIMONS

CREATIVE EDUCATION · CREATIVE PAPERBACKS

Published by Creative Education and Creative Paperbacks
P.O. Box 227, Mankato, Minnesota 56002
Creative Education and Creative Paperbacks are imprints
of The Creative Company
www.thecreativecompany.us

Design and production by Blue Design, Inc.
Art direction by Tom Morgan
Edited by Jill Kalz

Photographs by Photographs by Alamy Stock Photo/imageBROKER/
Mark Newman, 11; Dreamstime/Afhunta, 28, CrailsheimStudio, 3, 6, 29,
Crazy80frog, 18, Friedemeier, 26, Mbolina, 15, 16, Nikovfrmoto, 6, Ondrej
Prosicky, 21, Suzanna Ruby, 10; flickr/Biodiversity Heritage Library, 8, 10,
14, 16, 20, 22; Pexels/Leon Aschemann, 4–5; Shutterstock/Chris Alleau-
me, 9, Civdis, 29, Johannes Zoetekouw, 22, Sam DCruz, 21, warmer, 29;
Unsplash/Emmy Shingiro, 27, Jochen van Wylick, 23, Julie Wolpers, 29,
Petra Bouchalova, 24, Random Person, 17, Stephen Tafra, cover, 1; Wiki-
media Commons/Dominik Angstwurm, 14, Reinhard Onasch, 12

Library of Congress Cataloging-in-Publication Data
Names: Simons, Lisa M. Bolt, 1969- author.
Title: Hippopotamus / by Lisa M. Bolt Simons.
Description: Mankato, Minnesota : Creative Education, Creative Paper
backs, [2026] | Series: Spotlight on nature | Includes bibliographical
references and index. | Audience: Ages 10-13 | Audience: Grades 4-6 |
Summary: "Wallow in the hippopotamus's world with this nature title for
middle-grade wildlife lovers. Informational text pairs with a narrative
about a single hippo family to spotlight the wild animal's life cycle,
supported by infographics and a timeline of developmental milestones"—
Provided by publisher.
Identifiers: LCCN 2024048397 (print) | LCCN 2024048398 (ebook) |
ISBN 9798889896210 (library binding) | ISBN 9781682777879
(paperback) | ISBN 9798889897019 (ebook)
Subjects: LCSH: Hippopotamus—Juvenile literature.
Classification: LCC QL737.U57 S56 2026 (print) | LCC QL737.U57
(ebook) | DDC 599.63/5—dc23/eng/20241207
LC record available at https://lccn.loc.gov/2024048397
LC ebook record available at https://lccn.loc.gov/2024048398

Printed in India

CONTENTS

MEET THE FAMILY

COMMON RIVER HIPPOPOTAMUSES

of Murchison Falls National Park

Murchison Falls National Park is one of 10 national parks in Uganda, a land-locked country in eastern Africa. Established in 1952, the park is the country's largest conservation area and one of the oldest.

Within the park, the White Nile River squeezes through a canyon of dark rock 20 feet (6.1 meters) wide. It then dives 148 feet (45 m) to form the Murchison waterfall. The park also has **savanna** and forest. More than 550 species of birds and 140 species of mammals, such as giraffes, elephants, leopards, and hippopotamuses, make their home there.

It's April, a month during the rainy season. A **bloat** of hippopotamuses is **submerged** where the Nile River turns from Lake Albert.

A female hippopotamus has waited eight months for this moment. She removed herself from the bloat about a week ago. Now, standing in the shallow water alone, her body tightens. New life is about to begin.

CLOSE-UP

Shut tight

From the time a hippopotamus is born, its ears and nostrils automatically close to keep out water. Often, the calf is born in water. Because the ears and nostrils close, the calf can drink its mother's milk underwater.

CHAPTER ONE

LIFE BEGINS

Hippopotamuses are large, barrel-shaped, grayish-brown mammals with short legs. They are semi-**aquatic**. They spend part of their lives in water and part of their lives on land. There are two species of hippopotamus, the common and the pygmy. Common hippos are about 5 feet (1.5 m) tall. Adult males average about 16 feet (4.9 m) long, while females are about 14 feet (4.3 m). They weigh an average of 3,200 pounds (1,452 kilograms). Pygmy hippos are smaller, standing half as tall and weighing half as much as common hippos.

Hippopotamuses have **adaptations** that fit perfectly in the semi-aquatic world. Their eyes, nostrils, and ears sit on top of their heads. They have webbed feet. Their bones are very dense, allowing hippos to walk in deep water without floating to the top. Their short legs balance their big bodies on land when it's time to graze. Hippos must have access to water to keep their skin moist. If the skin dries out in the hot sun, it will

COMMON RIVER HIPPOPOTAMUS MILESTONES

DAY 1

- Born in water or on land
- Back legs come out first
- Nurses underwater
- Weight: almost 100 pounds (45 kg)

crack. Hippos can survive for a while in muddy areas if water isn't available.

After spending the day in the water, hippopotamuses lumber on land at night to eat. Hippos are primarily herbivores. They eat mostly grass and other plants near the water, although they have been known to eat meat on occasion. The animals rarely go farther than 2 miles (3.2 kilometers) from water to find food. Hippopotamuses grasp vegetation with their lips and tear it. They chew with their back teeth, called molars. Their stomachs have three chambers that can hold two days' worth of food. Each day, hippos eat between 55 and 88 pounds (25–40 kg) of food. They could go weeks without eating, if necessary.

CLOSE-UP

Jaws

The hippopotamus's jaws are huge. They open to nearly 180 degrees. The animal's bite is three times stronger than a lion and can cut a human body in half.

FEATURED FAMILY

Welcome to the World

The female hippo gives birth in the shallow water of Murchison Falls National Park. Her calf, a miniature version of its mother, enters the water back legs first. It awkwardly tries to swim in its new watery world but isn't finding its way to the surface. The mother dives underwater. With her snout, she nudges her calf upward. The calf pops above the water and shakes its head, ears flapping. Its nostrils open to take a first breath. The mother swims close, and the two immediately begin to bond.

Hippopotamuses are native to Africa. Common hippos live in countries south of the Sahara Desert, while pygmy hippos live in Liberia, Ivory Coast, Guinea, and Sierra Leone.

Hippopotamuses are one of Africa's most aggressive and dangerous animals. But they all start out as helpless babies, called calves. A newborn calf doesn't have teeth. It feeds solely on its mother's milk for the first month of life. It also has no way to protect itself. It relies on its mother and other hippos to keep it safe.

1 HOUR

- Learns to swim
- Can hold breath for about 40 seconds

1 MONTH

- Starts to graze

CLOSE-UP

Webbed toes

The common hippopotamus has four webbed toes on each foot. The webbing helps hippos walk in watering holes and muddy and rocky riverbeds. However, despite this feature, hippos cannot swim.

FEATURED FAMILY

First Meal

The calf returns underwater to nurse, **instinctively** finding one of its mother's teats. It nurses about 40 seconds and then goes to the surface to take more breaths. It does this over and over again—up and down, up and down. In the meantime, the mother sees a crocodile lurking on shore. She shifts her large body between the predator and her calf, ready to kill the crocodile if it moves too close. Minutes pass, and the toothy reptile wisely moves on. The calf stays safe by its mother's side.

The hippopotamus is the
THIRD-HEAVIEST
land mammal alive.

African elephant: 6,600–13,000 pounds (2,994–5,897 kg)
white rhino: 3,080–7,920 pounds (1,397–3,592 kg)
common hippo: 3,053–4,552 pounds (1,385–2,065 kg)

(2) MONTHS

- Nurses and grazes
- Can remain underwater about two minutes

CLOSE-UP

Canines

A hippopotamus has two sharp canines on the lower part of its mouth. These teeth can grow to almost 2 feet (61 centimeters) in length and are used mostly for fighting.

CHAPTER TWO

EARLY ADVENTURES

The mother hippopotamus and her calf stay separate from the bloat for 10 to 14 days after the birth. They form a bond. The calf learns its mother's voice and scent. After the two return to the bloat, other hippos help protect the calf. Hyenas, leopards, and lions are threats. For the first year of a calf's life, so are Nile crocodiles. Sometimes, even other hippos are dangers. Calves may get caught in the middle of fights between adults. They may be crushed or bitten, even by a parent.

Other animals aren't the only threat to hippopotamuses. People kill thousands of hippos each year. Some are killed in self-defense or to harvest meat. But many are hunted illegally for their ivory teeth, skulls, and skin. Orphaned calves, especially those still nursing, have little chance of survival without their mother.

There is safety in numbers, so bloats leave the water together to graze on land. A bloat varies in size from 2 to 150 hippos. The average is between

6 MONTHS

- Stops nursing
- Grazes with mother

CLOSE-UP

Eye shield

Hippopotamuses can spend five minutes at a time completely underwater. They have a clear **membrane** that covers and protects their eyes, much like having their own set of swim goggles.

FEATURED FAMILY

Time to Graze

After the first month, it's time for the calf to begin eating at night like the other hippos. It takes practice, though! The calf starts with a piece of bamboo. It tries and tries, but its lips can't quite grasp the hard round stalk. The calf gives up and nudges its mother. It will drink more of her milk once they are back in the water. In six to eight months, the calf will stop nursing as its teeth continue to grow.

10 and 15. The animals eat by themselves, but calves stay close to their mothers. After eating, the hippos may sleep on sandy ground, returning to the water in the morning.

The only close relationship in a bloat is between a mother and her offspring. A mother may be surrounded by up to four of her calves born back-to-back. They stay with her for years. Meanwhile, the youngest calf naps, plays, and nurses under her ever-watchful eye. It learns new skills, such as walking underwater. Calves grow quickly in size, especially the first few years.

(12) MONTHS

- Weighs 500 pounds (227 kg)

CLOSE-UP

Mine!

Hippopotamuses are very territorial. They want others to know where their land is and with whom they live. The lead, or dominant, male uses his long tail to flick his **dung** onto the ground and other hippos, marking it all as his.

FEATURED  FAMILY

Give It a Try

Night falls. All of the hippos in the bloat rise from the water and head down the well-worn path to the grassland. The calf's mother stands away from the others and begins to rip soft grass with her lips. The calf watches. It sniffs the grass and then mimics its mother, tugging at the short blades. The calf isn't as strong as its mother. Its head doesn't quite turn fast or hard enough to yank the grass well. But it tries again.

Hippopotamuses SLEEP underwater, rising to the surface to breathe without waking up.

7 YEARS

- Males start jaw-to-jaw fighting
- Males start "yawning," or threat, contests

CLOSE-UP

On top

A hippopotamus's ears and bulging eyes are on top of its head. Its nostrils are on top of its muzzle. Because of these features, the hippo can breathe while its entire body is submerged. It can also keep an eye on its surroundings.

CHAPTER THREE

LIFE LESSONS

A female common hippopotamus, or cow, is ready to mate when she is between 7 and 15 years old. A male hippo, or bull, is ready to mate a couple years earlier. A dominant male with a lot of land or territory will have the most success mating.

Male hippopotamuses are quite territorial when it comes to their mating rights. Bulls make what are called "dung middens," or piles of poop, at the paths and borders where their territories meet water. The bulls add to these dung middens every night when they graze. Males also show how defensive they are when they meet other males at the border of their territories. They whip dung at each other with their tails.

Mating between hippopotamuses usually happens during the dry season. A bull courts a cow by "wheeze-honking" at her. The two mate in the water, with the female completely submerged. Her pregnancy lasts eight months. When she gives birth, one calf is born. Twins are rare. The

7.5 YEARS

- Offspring leave their mothers

bull is generally not allowed near the calf. In the wild, hippos live for about 40 years. In captivity, they may live about 10 years longer.

Pygmy hippopotamuses are classified as endangered. There may be less than 3,000 left in their native habitat. Humans are cutting down or burning the pygmy hippos' habitat, and rivers where they swim are being polluted. The leopard is thought to be the pygmy hippo's only natural predator. Pygmy hippos are shyer than common hippos and would rather flee than fight.

Female pygmy hippopotamuses usually give birth to one calf after a six- to seven-month pregnancy. Because the calf can't walk well, the mother hides it in the bushes while she eats. Pygmy hippos often live alone or in small family groups. Their life expectancy is about 15 years shorter than that of the common hippo.

FEATURED FAMILY

This Is How It's Done

The calf, now six months old, is underwater drinking the last of its mother's milk. The sun is setting, and it's time for the bloat to graze. The calf pops up its head, takes a deep breath, and follows its mother out of the water. The hippos walk down the trampled path. The calf and its mother break from the bloat and find a lush area. The mother begins to eat. The calf grabs some grass with its lips, tears it by shaking its head back and forth, and grinds its meal with its still-growing molars.

CLOSE-UP

Blood sweat

Hippopotamuses **secrete** a blood-red "sweat" that isn't blood or sweat at all. It's a sunscreen, lotion, and germ-killing cream in one. The oily secretion helps keep hippos' skin protected when they're out of the water.

10 YEARS

- Females give birth to first calf

25 YEARS

- Females stop growing (males never stop growing)

CLOSE-UP

Vocalization

Hippopotamuses communicate with each other using bellows, snorts, grumbles, and "wheeze-honking." Some sounds can be as loud as 115 decibels, or the same volume as a jackhammer, and can carry more than 1 mile (1.6 km).

FEATURED FAMILY

Practice Makes Perfect

At four years old, the young hippo is fully grown. And it has two siblings! The mother stays close to all of them but pays the most attention to the calf that's still nursing. Just before sunset, the young hippo leads its mother, her new calf, and its two-year-old sibling out of the water. They head down the path with the rest of the bloat to the grassland. Once there, the young hippo distances itself in the dense vegetation. It tears off some grass and grinds it down as the sun slips below the horizon.

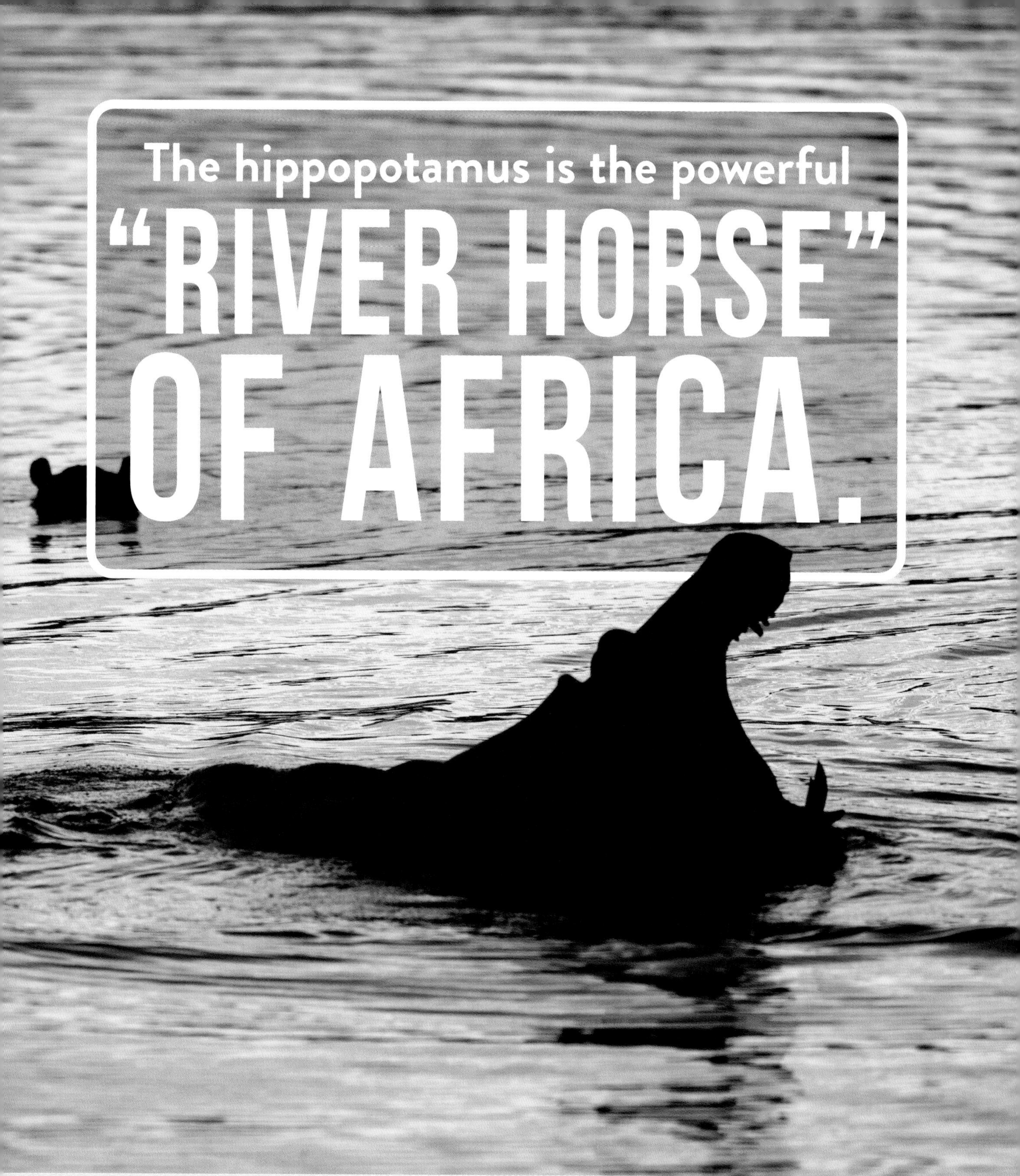

The hippopotamus is the powerful "RIVER HORSE" OF AFRICA.

(40) YEARS

- End of life in the wild

(50) YEARS

- End of life in captivity

CHAPTER FOUR

HELPING THE HIPPOPOTAMUS

Hippopotamuses are one of the world's deadliest animals. Each year, up to 500 people die from hippo attacks. Often, confrontations happen because hippos are territorial by nature. As people develop land and expand their farms and cities, they find themselves sharing living space with hippos. One response is to kill the animals. Hippos, however, are essential to their **ecosystem**. Their dung fertilizes vegetation, and their grazing keeps waterways open and flowing.

Many organizations are working toward a long, positive future for wild hippopotamuses. The Wechiau Community Hippo Sanctuary (WCHS) in Ghana wanted to not only protect its hippo population in its natural habitat but also recover its decreasing number. A shea butter processing center employs local people to provide income for the safe space. This collaboration benefits both people and animals.

The International Union for Conservation of Nature (IUCN) has a Hippo Specialist Group. Its goal is to find money to support research and conservation projects for both the common and pygmy hippopotamuses. This group wants to make sure both species are on the planet for a long time.

To fight the profitable illegal hunting business, England's Prince William and The Royal Foundation started United for Wildlife in 2013. This organization works with businesses around the world to protect endangered species such as the hippopotamus. The businesses help prevent illegal wildlife trade and take those who make money doing it to court.

The San Diego Zoo is part of an international wildlife partnership that works on conservation to support ecosystems around the world. During daylight hours, the zoo offers online visitors a streaming Hippo Cam to watch the facility's hippopotamuses.

There is still much to learn about hippopotamuses. A 2024 study from the Royal Veterinary College in England found that hippos become airborne when they trot! Studying the movement of the hippo helps large-mammal veterinarians with their understanding of the animals in their care. It also helps them diagnose and treat hippos when they have problems moving.

Studying the hippopotamus and finding ways to conserve its shrinking habitat will benefit the entire ecosystem. These efforts will protect the "river horse" of Africa for generations to come.

FAMILY ALBUM

SNAPSHOTS

The common hippopotamus is the most numerous subspecies. There are between 115,000 and 130,000 alive in the wild today.

Common hippopotamuses can trot up to 22 miles (35 km) per hour.

The algae that grow on a hippopotamus's waste are important fish food. They also carry **nutrients** into the waterways for other animals.

Hippopotamuses act like lawn mowers, helping to control vegetation around wetlands.

A bloat has both female and male hippopotamuses, but there is only one lead male.

The pygmy hippopotamus is the only other subspecies. Its skin color is greenish-black.

Pygmy hippopotamuses live mainly in forests and eat leaves, roots, and fallen fruit.

While common hippopotamuses have two or three pairs of incisor teeth, pygmy hippos have only one pair.

When a pygmy hippopotamus is born, it weighs about 13 pounds (5.9 kg).

Although its name means "river horse" in Greek, the hippopotamus isn't related to the horse. In fact, one of its closest living relatives is the whale.

WORDS to Know

adaptation	a change in a species to improve its chances of survival in its environment
aquatic	having to do with water
bloat	a group of hippopotamuses
dung	animal waste
ecosystem	a community of organisms and their environment
instinctively	automatically, without having been learned
membrane	a thin layer of skin
nutrient	a substance that promotes growth and maintains life
savanna	grassland with scattered trees
secrete	to give off or come out
submerge	to cover with water or put under water

LEARN MORE

Books

Gish, Melissa. *Hippopotamuses*. Mankato, Minn.: Creative Education and Creative Paperbacks, 2024.

Joubert, Dereck, and Beverly Joubert. *The Ultimate Book of African Animals*. Washington, D.C.: National Geographic Kids, 2021.

Roggio, Sarah. *Crocodiles vs. Hippos: Food Chain Fights*. Minneapolis: Lerner Publications, 2025.

Websites

"Hippo Cam." San Diego Zoo. https://zoo.sandiegozoo.org/cams/hippo-cam

"Hippo Fact Sheet." Nature: PBS. https://www.pbs.org/wnet/nature/blog/hippo-fact-sheet/

"Hippopotamus." National Geographic Kids. https://kids.nationalgeographic.com/animals/mammals/facts/hippopotamus

Documentaries

Adilardi, Ugo, dir. *The Secret Life of Hippos*. Sherman Oaks, Calif.: Slingshot, 2011.

Bestelink, Brad, dir. *Hippos: Africa's River Giants*. Arlington, Va.: PBS, 2020.

Duane, Graeme, dir. *Hippos after Dark*. Bethesda, Md.: Earth Touch, 2015.

Note: Every effort has been made to ensure that any websites listed above were active at the time of publication. However, because of the nature of the Internet, it is impossible to guarantee that these sites will remain active indefinitely or that their contents will not be altered.

Visit

CINCINNATI ZOO

Visit Fiona, Fritz, and the other hippos in their bloat.

3400 Vine Street
Cincinnati, OH 45220

PHILADELPHIA ZOO

Come see Unna and Cindy at America's first zoo.

3400 W. Girard Avenue
Philadelphia, PA 19104

SAN DIEGO ZOO

Catch up-close views (including underwater) of hippopotamuses.

2920 Zoo Drive
San Diego, CA 92101

WECHIAU COMMUNITY HIPPO SANCTUARY

This community-based area is home to one of the two hippo populations left in the country.

Weichiau Village, Ghana, Africa

INDEX